SO YOU WANT
—— TO BE ——
A DENTIST?

What You Must Know If You Want To Succeed In Dentistry

RYDER WALDRON D.D.S.
TROY STEVENS D.D.S.
MARCUS NEFF D.D.S.

ISBN: 978-1-4834-0212-3 (sc)
ISBN: 978-1-4834-0211-6 (e)

Lulu Publishing Services rev. date: 06/17/2013

CONTENTS

PREFACE

You're ready to be a dentist! It's an exciting time filled with anticipation, prestige, and opportunity. You've worked hard, sacrificed, and more than paid your dues. We are confident this book will help you hit the ground running!

We're three practicing dentists running general dental practices in the real world. We went to dental school and lived that experience together. Through the challenges of dental school we formed a tight bond.

We've remained close after school, texting each other and talking about dentistry, often on a daily basis. We've found that our experiences since dental school have been remarkably similar.

Over these several years, we've talked about how challenging dental practice is and how frustrating it can sometimes be. We've shared our experiences with challenging treatment plans. We've complained at times about how difficult it can be to be a solo dentist, a small business owner, and to succeed. With the benefit of hindsight, we know what makes the difference between success and failure . . . what can make the career of dentistry a joy or a regrettable pain.

Our intent in this book is to inform you—open your eyes, if you will—to the realities of practicing dentistry. We want you to enter this career with a full understanding of what you are committing to. After reading this book you will have a clear understanding of what you will experience each day and what it will take to be successful. You will be far more likely to avoid frustration and regret. Add in a passion for dentistry, and a dental career will reward you for a lifetime.

1

JOB OUTLOOK

At the time of publishing this book, there are sixty-four accredited dental schools in the United States. Of those, sixty-two have students enrolled. And of the schools that have students already enrolled, many are accepting more students into their programs than in previous years. According to the American Dental Association (ADA), there were 20,352 dental students enrolled in the 2010–11 academic year. That is the highest enrollment since 1980–81 when there were 22,842 students. In the last ten years enrollment has actually risen by an average of 1.7 percent annually. It is debatable if more dentists are needed. With the economy being in a recession in recent years, many dentists are working more years than they had planned. As in most industries, as people delay retirement, it makes it more difficult for new college graduates to find jobs. With dental school tuition increasing each year (some schools being over $300,000 for four years), it is more important than ever to find a job after dental school.

There is a growing need for dentists in rural areas of the country, but with smaller populations some graduates are hesitant to start or purchase a practice there. Another option is to work for a health-care clinic in rural America, but with tuition repayment being so high and salaries at these clinics not always being competitive with what can be earned in private practice, many dentists choose other options. However, some of these health-care clinics offer loan repayment incentives as part of the compensation package.

According to the ADA's website, in 2009 (the most recent study) there were 186,084 professionally active dentists in the United States. The US Bureau of Labor estimates that employment of dentists is expected to grow 21 percent from 2010 to 2020, faster than the average for all occupations. It also estimates that employment of dentists is not expected to keep pace with the increased demand for dental services. What it doesn't estimate is where these dentists will be employed, what their income will be, and if they will be working the amount of hours they would like.

One of the most encouraging facts is that on average people are living longer and keeping their teeth longer. In the past many people expected to have complete dentures at some point in their lives and were more likely to extract teeth that may have been savable with other treatments. As the public becomes more educated as to the importance of keeping their teeth, more demand for dental work is expected. Furthermore, people are becoming more aware of the link between oral health and overall health, which will give patients another recognized benefit of keeping their six-month recall appointments.

With advancements in dental technology, it is possible to keep teeth throughout a person's life. It makes early diagnosis possible. It also makes it possible to replace a person's missing teeth using dental implants. Dental implants are rapidly becoming more common. They make it possible to replace individual and multiple teeth. For those who are missing all of their teeth and are struggling with complete dentures, it is now possible to make implant-supported dentures or bridges. These implant options are improving the quality of life of many people with missing teeth, with many dentists placing implants and most restoring them.

As fluoride use becomes more prevalent, people will keep their teeth longer and longer. Most patients today have dental insurance coverage, which makes dental care more affordable. And with increasing demand for teeth whitening, veneers, and other cosmetic dental treatments, people may be more likely to use their discretionary income on dentistry. However, most trends show that dental office production has decreased across the country due to the slow economy. Many people struggle to afford food and a home, let alone dentistry, but hopefully the trend will reverse.

2

DENTAL EDUCATION

There are many prerequisite classes you will need to take prior to applying to dental school. For specifics you will need to contact each school you are interested in applying to. Most schools require the same classes, but there are a few that have different requirements. Most dental schools prefer a bachelor's degree prior to entrance. All dental schools require that you take the DAT (Dental Admissions Test). This test has sections in chemistry, math, perceptual ability, and other biology topics that have been deemed necessary for a student to know prior to school. There are many resources out there on how to prepare for this test, so this book will not go into those topics. We will say this, however: It's definitely a plus to have a high DAT score and a high GPA to get into dental school in today's competitive world. Dental school admission is becoming more and more competitive and therefore only those with the higher scores will be looked at. For example, there were 12,001 individuals who applied to the 2010 entering class of US dental schools. The number of first-time enrollees was 4,947(http://www.adea.org/publications/library/ADEAsurveysreports/Pages/ADEASurveyofUSDentalSchoolApplicantsandEnrollees20102011.aspx). So, in 2010, roughly 41 percent of applicants were accepted.

Nearly everything you need to know about dental education can be found at www.adea.org. For the application part you will need to fill out the AADSAS (Associated American Dental Schools Association Service, https://portal.aadsasweb.org). This is a standardized method of applying and getting your applications out to the schools you would like to attend. Plan on applying about one year before you actually want to enter dental school. The process takes time, and if you aren't careful, you will find yourself waiting a year because you missed deadlines.

Dental education is expensive. Actually, it is some of the most expensive education there is. Be prepared to pay top dollar for dental school. When all is said and done, you could buy a very nice home for the price you will pay for your dental education. When I came out of dental school, I was over $200,000 in student loan debt. (Let it be noted that I went to a private school and paid top dollar in tuition.) Dental education is increasing in cost every year; we have heard of graduates being in the $300,000–400,000 debt range upon graduation (more if you specialize). Unless you come from a very wealthy family (which we did not), that will be debt in the form of student loans. This link has great information on how to obtain financing for dental education: http://www.adea.org/publications/Documents/OG_2011/Chapter%204.pdf.

You must consider the cost of education in your long-term financial planning. We were under the assumption that the cost of education would easily be paid by what we would earn as dentists. The cost of education is a heavy burden that will haunt you for years and years if you don't plan for it. This becomes particularly important when and if you buy a dental practice. We will go more into the cost of ownership in a later chapter.

We will say this, though: We really enjoyed dental school. We felt like it was some of the best education that we received. For the most part the topics were interesting and we felt that the instructors gave us a really good foundation to start in the real world of dentistry. We have heard of some who have not had such a great experience in school; therefore, we must believe that it's school-dependent. It clearly is in your best interest to visit each school before you make the decision where to attend.

Different dental schools have different strengths. You will want to do your due diligence in finding out what schools are particularly good in areas that interest you. For example, our school had a graduate orthodontics and endodontics program. Therefore, we did not get a lot of experience in those disciplines clinically because those cases were sent to the postdoctorate specialty programs. This is good if you are considering orthodontics or endodontics, as you can get to know the faculty and spend extra time in those clinics getting known. This can help your chances of being accepted into those specialty programs. We, however, did not have interest in specializing and wanted to be general practitioners. We found that our school was very good in teaching general dentistry. We felt that when we graduated we were very much safe beginners. If you are interested in specializing, then you will want to look at your school carefully and ask questions prior to entrance so you know what you are getting yourself into.

3

PRACTICE OPTIONS

What is great about this book is that we, the authors, have all followed different career paths with different practice models, so we have a variety of practice experience. One or all of us have worked in different offices as associates, started practices from scratch, worked in a partnership, purchased a practice, worked as a solo practitioner, employed associates, rented office space, and built a new office building from the ground up. We know that after graduation there are several options for a new dentist, which include, but are not limited to: working as an associate, purchasing a solo practice, buying into a group or partnership practice, going to specialty school, doing a general practice residency, working in the public health-care system or prison system, education/teaching, starting a practice from scratch, research, and/or product invention. We will discuss the advantages and disadvantages of some of these options in order to help you decide which path may suit you.

Ask yourself this question: "Why do I want to be a dentist?" Is it so you can be your own boss? So you can make a lot of money? So you can work when you want? Because you're a "people person" and/ or "good with your hands"? Is it because you want to be able to choose whom you work with? So you can choose where you work? These are all questions that need to be answered honestly before deciding which career path to follow.

SOLO PRACTICE

According to a study released in 2010 by the American Dental Association, 92.3 percent of all professionally active dentists are private practitioners. Of these private practitioners, 58.9 percent are sole proprietor practice owners. There are many benefits to being a solo private practice dentist. Many predental and dental students want to be dentists because they want to own and run their own business. They want to be able to create their own vision for their practice. This is absolutely possible.

As the sole owner of your practice, you are afforded the option of working the days you want, as well as what hours you want to work on those days. After eight years of college education with a schedule that had very little flexibility, you can now choose your own schedule. Most dentists seem to choose to work 8 to 5 Monday to Thursday, but in today's ultracompetitive environment that is quickly changing. Many dentists are having to work nights and even weekends to accommodate their patients' busy lives.

Another advantage of being a solo practitioner is that you can choose who works for you. It can be very stressful trying to find the right people with the necessary experience. Your employees need to be able to mesh with your personality as well as with the other employees. It only takes one negative attitude to drag down the entire staff. We usually follow the rule that we can teach our employees how to do most of the dental procedures they will be required to do, but we cannot teach them how to have a warm and welcoming personality that is needed in a dental office. Team members with inviting, optimistic, and enthusiastic personalities can make patients feel more comfortable coming back to your office as well as make your days go smoother. Those who are usually down, with no passion for dentistry, can make you and your patients dread coming to your office.

Most dentists have certain procedures they enjoy doing and some they don't. Being your own employer, you are able to choose which treatments you will offer. If you don't enjoy extractions, refer patients needing extractions to an oral surgeon. If you want to learn how to place implants, you can take continuing education courses to learn how. In most areas of the country there are skilled specialists that are within driving distance.

With many of the benefits of being a solo dentist and practice owner come some drawbacks. While it is nice to be your own boss, it is also often stressful. You will be the one who must praise your staff, discipline for poor attitude or performance, give reviews and raises, help resolve disputes between staff members, and hire and fire. These actions are not always easy to take, especially since you must work closely with your staff all day. It can be uncomfortable at times.

A solo dentist practice is the dental career that most dental students aspire to. If you enjoy running a small business and managing all that that entails while also doing dentistry, this may be the career choice for you.

ASSOCIATESHIP

For many dentists, solo practice ownership is not attractive. They don't want the stress and hassle of dividing their focus between running a small business and performing the dentistry. As current practice owners, we have learned that working as an associate does reduce stress in some areas but creates it in others. It is great to be able to get to the office each morning and solely focus on doing dentistry. There are no business bills to pay, no staff to manage, no equipment to purchase, no practice goals to set, and no co-payments to collect. The owner dentist should do all of this. As an associate it is much easier to leave the office at the office.

Like other dental careers, though, there are some drawbacks to associateships. Usually an associate dentist must work with the staff, equipment, and supplies that the office currently employs. You will not get to have much, if any, input in staff hiring and firing. If you don't enjoy working with the office's assistants, hygienists, front office staff, or possibly the other dentist(s), it can make for long days. Most offices are already equipped before hiring an associate dentist, so you may have to adjust to using chairs, X-rays, and hand pieces that you aren't familiar with. This can be difficult but can be overcome with some effort. Furthermore, most offices order supplies (such as bonds, cements, and instruments) that the owner dentist likes to use and for which he approves of the cost. This may be the area in which an associate dentist can be accommodated.

Most dental offices in which you may work as an associate will already have an established patient base. As such, most of the patients have come to expect certain procedures to be performed at this office. For example, if the owner dentist likes all root canals to be completed in his office, you may feel some pressure to do all root canals, including molars, even if you don't enjoy or don't feel comfortable doing them. This doesn't mean that you can't refer patients to specialists, but you may feel as if you are expected to attempt some cases that you normally wouldn't.

Finally, as an associate dentist you will most likely not have a lot of input into the overall direction of the practice, whether that be which insurances will be accepted, marketing ideas, targeted patient bases, or treatments offered.

An associateship is a good way to start a dental career. It gives you a chance to be employed at different offices simultaneously and experience the ways in which those offices operate. It will allow you to try different types of equipment and technology, get comfortable with dentistry without being concerned with running a business, and get advice from different dentists. Although, as stated above, you will most likely not have a lot of input into the overall direction of the practice, you will be able to concentrate on doing dentistry without the stress of managing a small business. Also, as an associate you will have many more options to work in various locations throughout your career, because you won't be tied down as an owner of a practice. This is the reason many dentists choose to work as associates throughout their careers.

PARTNERSHIP

Another possibility is to buy into a partnership practice. Usually a partnership is equally owned by any number of partners, but most commonly it is two partners. The main benefit of being in a partnership is that you can share the practice responsibilities. It doesn't fall on you alone to make decisions regarding hiring and firing, equipment purchases, practice systems, and goals of the practice. You will have a partner to consult when you are contemplating a new hire or a new piece of equipment, for example. This can also be a disadvantage to being in a partnership. Basically, you are sharing the practice's money and, therefore, will have to decide with your partner(s) when or if you should buy a new piece of equipment. Also, one of you may like a certain employee and one may want to find someone to replace that employee. Who gets to decide?

Another potential problem with a partnership is how much you each get paid. Some partnerships pay each partner equal salaries each month. Some pay a percentage of production for the month. If you are getting paid equal amounts and one partner actually produces more than the other(s), will he or she feel shortchanged? If you get paid on percentage, will there be competition for more productive cases? It can be helpful if one partner enjoys doing cases or treatments that the other does not, such as root canals or extractions. This can keep as much treatment as possible in-house.

If you decide to buy into a partnership, it is vital that you and your partner(s) have nearly the same practice philosophies so as to avoid as many disagreements as possible. There will occasionally be tension in any partnership, but it is lessened by a harmonious relationship among partners. A partnership is a good choice if you want to own a practice but also have someone to lean on for big decisions. You will most likely have the same opportunities to do dentistry as in a solo practice or associateship but with the business stress decreased.

While most dentists work as a sole practitioner, as an associate, or in a partnership, there are other job options in dentistry that include, but aren't limited to, teaching, community health care, working in a jail or prison, and research and invention. This is why you need to be honest with yourself about why you want to be a dentist. Some dentists would prefer being a professor at a dental school and not having to see patients on a regular basis or run a practice. Some are concerned about lack of dental care in rural areas, so they may choose to work in a health clinic or volunteer their time and treatment. There are risks and rewards to each choice. There isn't one perfect option for all dentists, so deciding which factors are most important to you will help guide your decision.

4

STAFFING

Staffing is a necessity. In order to provide care in an efficient and cost-effective manner, you need to have someone to help you. You simply cannot be answering the phones, scheduling patients, posting payments, determining treatment planning, setting up rooms, turning over rooms, suctioning, and sterilizing instruments all by yourself. You need a team to support you in these tasks. With staff comes an entirely different dynamic to the dental office. Usually you have someone at the front desk who answers phones, schedules procedures, and takes payments. You have a dental assistant who can help in ordering supplies, perform chair-side duties, seat patients, and the like, as well as a dental hygienist who helps maintain the oral health care of your patients by cleaning, scaling, and root planing, and in a few select states the dental hygienist can also help you with administering anesthetic or even placing fillings. Without these staff members we would not be able to do our jobs efficiently.

When the office is running in perfect harmony and everyone is getting along, it is wonderful. On the other hand, when disharmony occurs and if someone on the staff is not pulling his or her weight or is disgruntled with another employee, look out! This is the day that will not go well. It is extremely difficult to be happy and give your patients the proper care they deserve when there is utter turmoil going on in the background. In our offices we have spent thousands of dollars to try to create a team that works well together, is on the same page, and has the same goals for patient care. This is one of the biggest struggles in dentistry, keeping the team harmony! I have heard it once put that a dental office is like a bus. You need to have the right people on the bus to make it to your destination, and you also need the people on the bus in the right seats to be highly productive and effective. We have had to shuffle people around before in the office because we found that their strengths were in a different area and we could utilize them better in that capacity.

It is very difficult to keep all the staff trained in all areas at all times. Team turnover leads to challenges that require training new team members, and that can be troublesome. Team turnover can happen for a few reasons. The most common reasons we see are an employee moving and the rare need to let someone go. We have found that when the need arises to let someone go—due to poor performance or for causing animosity in the office, for example—it must be done quickly. The employee who is not performing to his or her ability or is not meeting the needs of the office can drag down the morale of the team. It is never a fun experience to let a staff member go; however, this must be done. After the discomfort of doing it is over, you will see improvement in the other staff members and the overall mood of the office will be lifted.

Bringing new team members in to fill vacancies gives new life to the office. Each person who is a part of the team has unique strengths, and when a new staff member comes in, it is a chance for office rejuvenation. Use these opportunities to train and improve your team overall. You need to develop a job description so that each team member is perfectly clear on what his or her responsibilities are. If you do not do this, you will find that the staff will say things like, "That's not in my job description." It's important to be clear on what your expectations of your employees are so they can do their best. To reiterate, if an employee doesn't follow the job description after he or she has received that description, make the changes necessary so the rest of the team does not suffer.

5

PATIENTS

Obviously the focal point of any practice is the patients. While every patient base differs due to location of the practice, the treatments offered, and other variables, some patient behaviors are universal. There are also some essential recommendations we will make to better serve your patients and to help you build a busier practice. Through almost thirty years of combined practice experience, we have learned some ways that make patients more comfortable at your office and more likely to refer other patients to you.

The most important recommendation we can make is to be honest with your patients. If you fill a severely decayed tooth and you think it may be painful after the anesthesia wears off, and may even require a root canal in the future, tell the patient. Give the patients all their treatment options even if that entails referring them to a specialist for some treatment. Your honesty will build trust in the patient-dentist relationship. Some conversations are difficult to have with patients, but patients will appreciate your willingness to tell them the truth.

Patients often appreciate a phone call after treatment to see how they are feeling. This is especially important with procedures that can be more painful, like extractions and root canals. I don't think patients want you to call them to see if they're recovered from their prophy. But they will be surprised and happy that you call to check on them after major work. It may sound like a waste of time, but if you are trying to give your patients more personal treatment and set yourself apart from other dentists, it is very important. The patients will thank you, and they will likely tell their friends and family how great their dentist is.

Not very often do dentists and doctors stay on schedule throughout the day so that all of their appointments start at the pre-appointed time. Make a point of staying on schedule in your office. People will be pleasantly surprised. Many of them will thank you, and you will feel much more relaxed. Everyone gets behind once in a while, but don't make it a habit or even a regular occurrence. Schedule enough time to get each planned procedure done. Don't put unnecessary pressure on yourself.

Oftentimes, patients will try to pressure you into doing treatment you aren't comfortable doing. For example, if you diagnose a root canal on a tooth with a calcified canal and you feel an endodontist could better treat the patient, then make the referral. Some patients may complain that they don't want to go to another office or that specialists are too expensive. Try explaining to them that you want the best treatment for them and you would feel better about them seeing a specialist. They may be unhappy that they have to make an appointment somewhere else and try to convince you that they trust you and know you can do it. They may even threaten to leave your office and find another dentist. A lost patient is better than doing treatment you don't feel able to do.

Patients may also pressure you to do is prescribe unnecessary medications. Drug-seeking patients are very smart. They know just the right words to say to appeal to your sympathetic side. They know that if they call you at home at night you are more likely to call in a prescription for them than actually take the time to go to the office. When we first began practicing, we would get calls all the time from patients who claimed they wanted to join our practices and would schedule an appointment as soon as they could, but for now they needed some pain meds to get through the night. A good rule is to never prescribe to anyone who isn't an established patient. If you do choose to go in after-hours to see a patient, make sure you always have an adult staff member or family member go with you. It protects you and it protects the patient. If you don't feel comfortable with the situation, don't go in. Better safe than sorry.

Being polite to patients and telling them thanks for coming to your office goes a long way to establishing a long-term relationship. Most people are nervous to go to the dental office. How you treated them and how much the procedure hurt are the main criteria patients will use to evaluate you as a dentist. Most people can't determine if you do a good filling, but they do know if it hurt and if you were sympathetic to their situation. You can be the best clinician in the world, but if your chair-side manner is cold and intimidating, patients may choose to go elsewhere.

6

CLINICAL CARE

Seeing patients will make up a huge part of your life as a dentist. Becoming really proficient at clinical dentistry is critical to your success. If you find there isn't much of dentistry you like to do, if you find it difficult to perform well—even with training and practice—you should strongly consider doing something other than being a dentist. At the very least you should try to limit your practice to doing those clinical procedures you enjoy and are great at. Don't settle for mediocrity!

There are so many amazing materials, techniques, pieces of technology, and ways to perform day-to-day dentistry. There has never been a better time to be able to provide excellent care to our patients. We are able to treat patients comfortably and quickly. We are able to complete a treatment that not only functions well but also looks natural. Additionally, today's materials hold up very well in the challenging oral environment.

The great news is that all of these materials, techniques, and technologies are advancing at a fast pace. Performing great dentistry will likely be easier next year than it is today! These advances also present more opportunities to be productive and successful.

The challenge will be staying on top of it. To stay at the forefront will require a conscious commitment of time and money. You will have to make plans and set goals. Budget for continuing education. You will have to really understand the business of dentistry to ensure that the money you're investing in being a top-notch clinician translates into growing profits.

Clinical dentistry requires intense focus. The mouth is a small, dark, and dynamic environment. As a dentist you are tasked with performing microsurgery and doing it exceedingly well. In most instances this microsurgery is performed on a fully alert human being with a whole host of emotions and physical limitations. Maintaining focus is integral to clinical success, but it is very taxing. Maintaining focus is very challenging, given the variety of patients you will see and the activities happening in the office all around you.

Clinical care can be highly demanding. It is not unusual to feel physically and emotionally spent at the end of a busy day of seeing and treating patients. It is important that you establish a daily routine of physical exercise so that you can consistently perform at a high level. You may not think of dentistry as an athletic endeavor, but the more physically fit you are, the better you will perform clinical dentistry. The more physically fit you are, the better you can recover from a challenging day in the office.

Just as important is staying emotionally fit. Day in and day out you are interacting with patients who range from happy to see you to those dreading the moment you show your face in the room. You will have team members who are supportive and others who are simply there for the paycheck. Some procedures will turn out wonderfully. Other procedures you'll get done and hope for the best. Still other treatments will not run smoothly, and you may have to make a referral or change the plan completely. As you surely appreciate, a day of clinical dentistry at the office, even the best day, can be a roller coaster of challenges and a mixture of emotions.

Find healthy ways to address your emotional needs. Avoid alcohol or drugs as an escape. They might offer momentary relief, but they will destroy your career and your life. Don't engage in activities that in the moment might feel good and give you a quick feeling of freedom but will cause problems later on. Instead, find hobbies and activities that build you up in other areas of your life. It may be gardening, jogging, or reading a good book. Maybe it's going to the gym. We even encourage you to look to spiritual pursuits like regular church attendance and prayer. All of these activities can provide you with a sense of purpose, an emotional release, and some much-needed perspective.

Be committed to great clinical care. Your patients often don't know whether the treatment you provided is good or bad. Aside from anterior cases, the patient's only gauge of your quality of care is: Did he hurt me? Do I hurt after? Can I function? Don't let the patient's inability to evaluate your clinical quality tempt you into cutting corners. You won't sleep well at night! It will catch up with you! That restoration will fail. You will have to redo the treatment and relive the stress of failed dentistry. The patient may go somewhere else, and you don't know how the other dentist will talk about you. What the other dentist says may encourage a patient to pursue legal action if your treatment was questionable. Just commit to doing excellent dentistry, and when processes don't or can't turn out well, be up-front with the patient.

On the same note, we all have seen a patient whose dental restorations look horrible. Avoid the temptation to badmouth the other guy! For one, you don't know under what challenging circumstances the dentistry was done. You don't know if the patient has done what's necessary to care for his or her mouth before coming to see you. By putting down the other dentist, you may invite problems for him that really aren't fair. If you see truly questionable dentistry from another dentist, have the decency to contact the other dentist and express your concerns. Be honest with the patient about his or her condition, but don't automatically blame the other guy, thinking it will put you on a pedestal in the patient's eyes. In reality you only harm the very profession you practice. Patients start to see all dentists as hacks barely worthy of their title, *Doctor*.

Know when to refer. It is tempting to try to be a one-stop dental shop. As a general dentist, there is nothing that stops you from providing any and all dental services. This can be a great opportunity! The more procedures you can perform, the more money you can potentially produce, and the more the wealth you can accumulate. However, remember that every specialty procedure you perform is held to a very high standard of care. There are recognized dental specialties because some procedures require such a level of precision and skill that someone would need special knowledge and additional training to do them correctly.

That's not to say that a general dentist should never do a root canal, extract a tooth, move teeth, or place an implant. Just take the time to really know what you're doing and that you can do it at the level of a specialist. Don't just take a weekend course on endodontics sponsored by a dental manufacturer and think you can do any root canal. Do what is necessary to really be efficient and excellent at the specialty procedure you are going to do.

As mentioned previously, it is a disservice to the patient to perform a procedure to an average level. It is especially egregious when a procedure could be done excellently if treated by the right clinician. Don't ever let money be your guiding factor when deciding what treatments to provide patients in your practice. It will come back to haunt you!

Clinical care is the bread and butter of dentistry. Commit to doing it at the highest level. Spend the extra time to get really good at it. If you do, you will enjoy greater success and peace of mind than the guy who just does what it takes to push the patients through.

7

DENTAL INSURANCE

This topic is the source of much animosity in the dental community right now. I personally feel that depending on how dentistry as a profession deals with the insurance issue will determine whether we follow the medical model (all practices owned by large corporations) or if the solo practice can survive. Over the years the cost to provide dentistry has gone up drastically. Dental supplies continue to be more expensive, as do equipment and new technologies. Dental insurance, on the other hand, has not increased dental benefits to keep on par with increasing costs of providing care. Therefore, the patient has had to pay more out-of-pocket. This is a difficult subject to discuss, as there are many views of it. As an employer you are looking to find a dental benefit package that is inexpensive (or at least manageable) so you can offer that to your employees as a "benefit." As a patient, you want to go to the dentist who is a "provider" for your insurance in order to get the most bang for your buck. So who does this hurt? The provider *and* the patient! The provider takes a hit in his or her fee. The patient, depending on the type of insurance he or she has, may not get to have as much doctor time as he or she would like or see his or her favorite doctor. We will explain this a bit later.

Dental insurance comes in different shapes and sizes. There are regular indemnity dental insurance plans. These plans allow people to pick their own dentist, and then they (the insurance companies) have a contract with the dentist to provide services for those people. The insurance company sets what their reimbursement will be, and the dentist is able to charge the difference to the patient. These plans are not terrible, but with insurance not keeping up with the benefit-to-cost ratio, it is often a bit more burdensome on the patient to pay the additional money owed. In a down economy it is particularly difficult, as times are tough for people. Patients will sometimes elect not do treatment because of the out-of-pocket expense.

The other type of insurance is a PPO (Preferred Provider Organization). With these types of insurance plans the provider must agree to write off and not charge the patient for any difference in cost. This is much better for the patient and much worse for the provider. And to make the story even better, you as the provider have no control over the fee of the procedure! Oh yes, you heard me right; the insurance company tells *you* how much they will pay you for your service. Ouch! This could cause the end of the solo practice in the long term. Let's do some simple math. Let's say you charge one hundred dollars for a procedure. That's the fee that you set based on your overhead and operating costs, with your profit margin built in. So you sign up for a PPO insurance because you need to get more patients in the door, right? Well, this all sounds nice and dandy until you see that for your normal charge of one hundred for the procedure, the insurance company contract says that they will pay you eighty-five dollars—and you *cannot* bill the patient for the remaining fifteen dollars. Let's calculate how much this is really affecting the bottom line. This assumes that you collect all the money that you actually billed out.

Normal Scenario (Cash)	PPO Scenario
$100 Procedure - $70 Overhead	$85 Procedure (this number is dictated by the insurance company, you have no control) -$70 Overhead
$30 Profit (before tax)	$15 Profit (before tax)

Can you see the problem with this scenario? By accepting this insurance you literally cut your profit by 50 percent! The problem with this is that all your debts and costs associated with doing the procedures get paid. Your staff gets paid, your lab gets paid, all the utilities get paid, and your dental supply company gets paid, but you get 50 percent reduction in your pay. This is the future! This is why we stated that how dentistry handles this very situation will determine if solo practitioners are able to survive. When we as providers lose control of the ability to charge what we need to cover costs and to get paid, we have given up too much autonomy. This is where dentistry is headed. You may ask why anyone in his or her right mind would sign up for such a horrible insurance. It's simple—to get patients. If you have empty chair time in the office, you want to fill those chairs however you can. It's a desperate measure, but something is better than nothing.

There are a few ways to help buffer the hit of the PPO reimbursement fees. You can negotiate with labs and dental supply companies, use different materials for PPO patients, and so on. Another difficulty we are seeing is that in order to make it profitable, you need to see more patients to get the same amount of money. This causes you to see twice the number of patients to get the thirty-dollar-profit you used to get by only seeing one patient. Cutting down on patient time is a double-edged sword. Some patients like to talk; others don't mind being rushed through. It's a fine balance that is patient-specific. We discussed some of this in the "Patients" chapter. At this time, with PPO patients, unfortunately we must change the process, or we will struggle to remain profitable.

Now, as mentioned earlier, there are different types of insurances and there are even more insurance companies. Each insurance company will have their set of specifics regarding the services that they will cover and at what percentage they will cover them. This leads to another challenge of ownership: paying a staff member to handle all the insurance questions and filings. You will deal with this daily. "Why did my insurance not cover this?" "I thought my insurance was going to pay this much and it only covered this much!" And the list of questions that you receive goes on and on. It is important to know in advance that it takes time and money to process insurance claims. Some insurance companies are very difficult to deal with; others are not so bad. You can have a staff member on the phone for thirty to forty minutes with one specific insurance issue—and you are paying for that. Yes, it is a benefit for your patient, but definitely something that you will want to keep an eye on in your practice.

8

TAXES

It has been said that nothing is certain but death and taxes. It is true. Someday you will die, and taxes are never going away. It is paramount to your success that you deal with taxes, both personal and business, proactively. Wherever you stand on the political issue of taxes, you owe it to yourself to clearly understand how taxes affect your bottom line and the success of your practice.

In starting a practice, one of your first decisions will be what type of entity or corporation to establish. It isn't within the purview of this book to advise you on what type of entity you should structure for your practice, but you will need to practice as something other than a private individual. You will need to be structured as some type of corporation. The most common corporate structures are limited liability companies (LLC), sub chapter S corporations (S Corps), and sub chapter C corporations (C Corps). Again, our intent is not to tell you what type of corporation your dental practice should be but to make sure you understand that the type of corporation you establish will affect your tax situation.

Depending on your corporation type, profits from the company may pass through to you and be taxed as personal income. The profits under other corporate structures would be taxed differently. Each corporate entity has its own distinct advantages and disadvantages. We recommend you discuss your situation with an accountant and an attorney to choose the entity that is most appropriate for you. The bottom line is that you don't want to create a tax burden when one could be avoided by simply practicing under a different corporate entity.

Taxes are a significant expense that you must budget for. If you wait until the taxman sends you a bill, let's just say it won't be a pleasant surprise. Most tax amounts can be anticipated with some budgeting and planning. Your business taxes will likely include the following: property tax (personal and building), payroll taxes (Medicare, Social Security, and federal withholding), sales and use tax, and unemployment insurance tax. As of January 1, 2013 there will also be an excise tax that is incorporated into the Affordable Health Act ("Obamacare"). There is still some discussion as to how this will affect dentistry, but it looks like dental devices will be considered medical devices and will be taxed.

These taxes can vary depending on the success of your practice, but with planning and budgeting and help from your accountant, you should be able to have a good idea of how much they will be. You will be prepared for tax payments instead of scrambling to figure out where the money is going to come from to pay them.

Don't blow off these tax payments. The government is a slow-moving machine, but once it is aware that you haven't paid a tax, you will be under relentless pressure to pay up. The government will get paid, and if they don't, you will have a hard time staying in business. If you won't be able to make a tax payment, get ahead of the problem and be proactive. If contacted in advance, the IRS will likely let you work out something. But ignore them and the IRS will be a nightmare that doesn't go away easily.

There may be some who start a dental practice with the sole intent of serving the public and doing charitable work. That is a noble pursuit.

However, most dentists enter into dental practice with a combination of interests. Yes, they want to serve people, but they also want to earn a living. They want to make money!

There is nothing—absolutely nothing—wrong with trying to be financially successful in your dental practice! In fact, we would submit that it is irresponsible to not be as financially successful as possible! If you are just getting by in your practice, how can you possibly invest in new technologies that will improve patient care? How can you improve your knowledge and skills by attending the best continuing education courses if you can't afford to go to them?

If you are not financially successful in your practice, will you be able to hire people and create more jobs? By being highly successful, you will have more money to spend on vacations, a home, a boat, services from others, charitable organizations, and your retirement fund. Your high success contributes to the success and well-being of people all around you.

Part of being financially successful requires that you pay attention to your tax liabilities. We feel it is important that you enlist the services of a competent and successful accountant. This accountant will help you structure your practice and your finances in such a way as to meet your tax obligations but limit your tax liabilities to their absolute minimum.

It should be your goal to pay as little tax as you are obligated to pay. In no way should you cut corners or cheat your way out of paying taxes. But you should also never feel guilty about doing everything in your power to keep more of your money. You earned it. You know best how to spend it! You will do far more good spending your money on the the causes and the stuff that matter to you than the government will.

9

DEBT

D ebt is a very powerful tool. It is also a very powerful master. Handle debt with the utmost caution. It is like fire! It can help and sustain you. It can also destroy everything you have worked for. In today's world it may be highly unlikely that you can completely avoid debt, but if you can, do it!

A TRUE STORY—MARCUS NEFF, D.D.S.

When I first got married, my wife's uncle gave us a book as a wedding gift—a book! To top it off, it was a book called *Rich on Any Income*. Yes, a book on budgeting. What a wedding gift, right? I mean, we didn't need some stinking book. We wanted a toaster, decorations, a TV, anything but a book, and especially not a book on money!

Well, I have to admit—it was a great blessing. We read it and used it for the early part of our marriage. I was in school, my wife was working, and we made very little money. We did, however, have a budget. We were in control of our finances. We knew where our money was and what we planned to spend it on. Every dollar had an assigned task.

After a few years, it was necessary to "buy" a car. I say "buy" because we actually didn't *buy* anything. We got a loan so we could drive a car we didn't have enough money to pay for. Oh well, the payment still fit within our budget. Next, I needed to pay for my education, and I didn't have enough money for that, so we took out some cheap and easy student loans. Even better, those student loans didn't have a payment attached until I was done with school, so they didn't affect our budget!

Then came dental school. There was no scholarship large enough, no rich uncle waiting in the wings to pay, so where did I turn? Uncle Sam. Student loans were readily available and easy to get. The school was well equipped to get the loan to me so they could get the money in their pocket. To make life even better, the government was happy to loan me more money than was required to pay tuition because they knew I would be a busy student with a lifestyle to support—I couldn't possibly be expected to work to pay for my living expenses!

So I took out very large student loans for four years of dental school. I'm ashamed to admit the budget kind of went out the window at this time. I mean, we still balanced our checking account and we pretty much lived within our means, but we weren't really watching our financial situation carefully. It was going to be fine—in four short years I was going to be a rich dentist!

Coming out of school, I had approximately $250,000 in debt, primarily all from my education—student loans.

After school, it was time to practice dentistry. I chose to start up a practice from scratch. I certainly didn't have the money to do it. Where did I turn? The bank! *Show me the money!* And the lenders were lining up!

After building a new building, equipping the dental office with new equipment, and getting some "working capital," I had added almost an additional one million dollars' worth of business debt to my bottom line. That is a lot of money no matter where you're from! And it all had to be paid back! But I was going to be a rich dentist!

Payments on that kind of debt amount to thousands of dollars each month. Even if you are buying into an established practice, this kind of debt is a significant obligation to meet month after month. It makes it that much harder to produce enough in your practice to be able to pay yourself.

You also have to account for a home mortgage and probably a car payment or two. That's just being conservative. We wouldn't recommend it, but many people also finance other purchases with debt, like four-wheelers, landscaping, camp trailers, home furnishings, vacations, electronics, and dining out.

Very quickly, because of the profession of dentistry that you have chosen, you can be sitting in a pile of debt! That debt has to be paid back. It never sleeps. The interest on that debt grows day and night, weekday and weekend. It never takes a holiday. Those debt payments rob you of money that you could be investing in something of value. If you were able to take the money you pay out in debt payments and apply it to, say, a mutual fund, you could have a very sizable fortune on your hands upon retirement.

Is it possible to become a dentist and build or buy a practice without debt? Probably not. Then what's the point of warning you about debt if becoming a dentist without debt is almost impossible? So that you will be exceedingly cautious! You will probably have to use debt, but acquire as little as possible. Be prudent and frugal where you are able. If you can equip your office nicely with used equipment (and less debt), do it! If you can drive a nice used car and pay cash for it, do it!

Be creative. Don't just accept that debt is the only way to get it done. You could join the military and have dental school paid for. You could have the practice seller finance the practice purchase (at much better terms and rates than a bank). You could choose to practice as an associate. While you wouldn't have ownership in the practice, you wouldn't have a mountain of debt. Learn to do without and to live within a budget. It sounds old-fashioned, and it is. It sounds simplistic, but it is just simple common sense.

Carrying a load of debt is a burden that can quickly and easily suck the fun right out of dentistry. Debt can make life feel like you are a gerbil running on a wheel in a cage. The worst thing is it's a problem of your own making. Debt can and will dominate you if you are not extremely careful. Use it with extreme caution and wisdom, and you will enjoy peace of mind and prosperity. Be deceived and think debt is your ticket to peace and prosperity, and you will end up with neither peace nor prosperity. It's tough to be successful in life or in dentistry if you are a slave to debt.

10

WHY WE CHOSE TO BE DENTISTS

Each person who has applied to, gotten accepted by, and attended dental school has a unique story of his or her own journey. We would like to share ours. One of the stories was shared in the previous chapter, and the other two are in this chapter. Our hope is that you can relate to some of our stories and use them to make decisions that will help you in your career path.

EXPERIENCE #1

When I was young, I had a dream of one day being a doctor. My uncle was a doctor, and I was always under the assumption that he had it made. As I grew, the desire never left me, and as I met other doctors, I confirmed that that is exactly what I wanted to be when I grew up. So when the time came to go to college, I chose premed as my major. There was actually one time that I considered switching to another major because the science classes were tough, and I felt that I didn't have the opportunity to learn about anything else. I was simply too busy maximizing the dollars that I spent and only taking classes that counted toward my major. I didn't have the time or money to study subjects like business and accounting, which really interested me as well.

I received an associate of science degree and then applied to a medical technology program to complete my bachelor's degree. My thought process was that medical school was very difficult to get into, so I wanted to have a backup plan. I didn't want to end up with a bachelor's degree and be unemployed. I worked at a phone telemarketing company, a supermarket, and a warehouse at various times, all while attending school full-time, bringing some money in to attempt to keep student loans to a minimum. My schedule was school from 7:30 a.m. to 3:30 p.m., work from 4:00 p.m. to midnight, and studying when I had the opportunity. When all was said and done, I ended up with my bachelor's degree in medical laboratory science and was just under ten thousand dollars in debt due to student loans. I was pretty burned out simply because working and going to school was taking its toll on me. I had started to study for the MCAT; however, I was so tired of school that I decided to get a job and not worry about more school for a while.

What was nice about my choice in degrees was that after I had passed the medical technology board exam, I was able to find a job in medical technology at a local hospital. (For those who are unfamiliar with what medical technology is, it's basically a lab tech. We run tests on blood and body fluids to help doctors with their diagnoses.) I was excited to get to work in the field that I had studied in. This I did for two years—then reality set in. I began thinking again of my dream of being a doctor. I had pretty much maxed out my potential in the medical laboratory field and wasn't making nearly as much money as I wanted to (Does anyone?). All of my friends were making good money in fields such as engineering, computer science, marketing, and so on. However, I wanted to be a doctor. Well, at least up until I started working in the hospital. My first job was a graveyard shift. I was the only person in the lab at night, and I had the pleasure (if you can call it that) to call doctors at all hours of the night with panic values on their patients. This was the first time in my life that I had really seen the behind-the-scenes life of a medical doctor. I actually got to rub shoulders, for the first time, with doctors and see what their lives were like while at work, not outside of work. When I saw my uncle, I never saw him actually at work. I saw the outside life he lived and thought, *Yeah, that's what I want.* I was not excited about the prospect of being called in the middle of the night only to hear horrible news about my patients. I was not excited about making life-or-death decisions. I was not excited about working eighty hours a week.

It was at this time that life really hit me hard. What now? I had spent my entire life wanting to be a doctor, and now that I actually knew what a doctor did, I did not want to be a doctor. I was stuck. The experience that I gained in those two years was priceless in a couple of ways. I learned that I would never again complain about how much it costs to go to the doctor. The years of schooling and the knowledge and expertise that these men and women have in taking care of people are worth every penny that I pay them. The second lesson I learned was, *Thank goodness I didn't go to medical school!* I would have been utterly miserable, because when I had those two years to really see what it was like, I realized that it wasn't what I wanted to do after all.

I decided to look around a little and see what else was available for me. I was not going to be making the living that I wanted if I stayed working in the lab. I had studied science, and that's all I knew. I wanted a good work schedule, I wanted to help people, and I wanted to make a good living for my family. Now that I knew that medicine was not going to be my career path, I had to find something else that would fit my needs. It was at this time that I met a man who had moved close to where I lived. He was a dentist. Dentistry! I thought maybe that's what would fulfill what I wanted to do. I decided to do some job shadowing and talk with him. He had just recently graduated and was very excited about starting his new practice. I did a few hours of job shadowing and then decided. This was it. *I'm going to apply to dental school*, I thought. Long story short, I took the DAT (Dental Admissions Test), was accepted, and went to dental school.

Had I not taken the two years to work in the hospital lab, I may have been a medical doctor and not liked it at all. The hope for this book is not to scare you away from dentistry but rather to inform you of the trials and pressures that come with the job. This is information that job shadowing will never give you. Clearly, the topics discussed in this book are not unique to dentistry. Many businesses experience the same struggles and stresses that dentistry does. We hope that by reading this book, and reading our experiences, you will have a little better understanding of what to expect as you enter the career of dentistry.

EXPERIENCE #2

Before college I hadn't given much serious thought to what I wanted to do as a profession. I thought that I would be most happy working in health care. I considered going into physical therapy or becoming some kind of a doctor. I thought it would be great to be a doctor, because they were rich, had nice cars, huge houses, and all that. But I also thought that the time commitment of four years of undergraduate school and four years of medical school followed by a residency would be more than I wanted to do. Also, I assumed I would need to have close to a 4.0 grade point average in college to even consider going to medical school. However, I did feel like I had the potential and desire to do something that would make a difference in people's lives. After two years of college, earning my associate of science degree, it was time for me to make a decision. I decided that I would try to get into physical therapy school and started working toward that goal. But after researching the length of physical therapy grad school, expected student debt, job competition after school, beginning salaries, and job satisfaction, I was not quite as enthusiastic about my decision. I kept working toward it, though.

At this time I was working at a pharmaceutical company, packing and shipping vitamins and supplements. I would go to lectures in the mornings and labs in the afternoon and then work 4:00 p.m. to 9:00 p.m., Monday through Friday. The most important lessons I learned while working at this job for three years are that I strongly disliked meaningless manual labor, being told what my work schedule would be, taking orders from others, having someone else determine my wage, and getting very few benefits. Essentially, I didn't like being an employee—and I was bad at being an employee. I wanted to be the boss. I assumed that if I was in charge, processes would run more smoothly, the employee morale would be better, my stress level would be lower, I would be able to take time off when I wanted, and I would pick the best employees and get rid of the bad ones.

Prior to working at the pharmaceutical company I worked as a janitor for three and a half years. I disliked many of the same details about being a janitor as I did packing vitamins. I didn't like either of these jobs, but I believe they served a purpose, and I gained some valuable insight into what I wanted to do with my future. By this time I was motivated to find a career that would allow me to be independent professionally. I still wasn't sure that physical therapy would provide that for me.

One of my coworkers had made the decision to go to dental school, and occasionally we would discuss our chosen career paths. I had never seriously considered dental school until one day my coworker told me I should look into it. So I did. What I found was that my grades were definitely high enough to get into dental school, I had already passed many of the coursed required for dental school, a dentist makes a much higher salary than a physical therapist, there is more job freedom in dentistry, most dentists are private practitioners, many are self-employed, and the public holds very high esteem for dentists. The drawbacks were that I would have to go to one more year of graduate school than I would for physical therapy, I would have to move out of state for school, and my student debt would be very high. After much consideration and research, I made the decision to go to dental school.

EXPERIENCE #3

Like many people who become dentists, I had a dentist whom I admired growing up. He was my family dentist. He was always kind. His staff was friendly and fun to be around. Going to the dentist and talking to him over the years, I learned that he only worked three and a half days a week. I was told he lived in a big house. I would see a nice car in the staff parking area and figured (rightly) that it was his. My dentist was a golfer and spent much of his free time from spring through fall golfing. In fact, in the summer he rearranged his schedule to have more time away from the office to play golf.

My older sister went to dental hygiene school and upon graduation became a hygienist in this dentist's office. I got an even more "inside" look at his practice and lifestyle. Annually, he would take his team to continuing education courses out of town. If they met their production goals, which they always seemed to, he would take the entire team on a vacation—Mexico, the Caribbean—somewhere exotic and fun!

Watching all this from a distance made me want to have that same lifestyle, fun, and freedom! So in my mind, early on, it was settled. I was going to be a dentist.

As I finished high school and headed off to college, especially in the last year or two of my undergraduate program, I made it a point to check in with this dentist and arrange time to "job shadow" in his office. Every time I went to his office, he was very kind and easy to talk to. His team was always pleasant and friendly as well. This fun, kind, friendly environment again reaffirmed that dentistry was going to be a great career for me. I mean, after all, I was a fun and friendly person; this had to be the right career!

You may have already noticed I haven't said a word about the actual dentistry. I haven't said a word about staff meetings, team training, and the like. Well, I never actually saw or experienced those aspects of the job. Sure, I was in the office, and I would watch the procedures. But it was always from a distance. I wasn't giving the shots. I wasn't using the drill. I wasn't placing the fillings, getting the impressions, extracting teeth, and all the other day-to-day steps a general dentist takes. At the end of the day, I went home. I didn't see the patients again. I didn't make post-op phone calls. I didn't work through staff issues.

So from the safe distance of an observer, dentistry looked, well . . . easy! I even remember thinking, *How hard could it be to inject some anesthetic, drill away part of a tooth, and then pack some filling in there?* I know it was incredibly naive to think that way. But how could I really know? There is no apprenticeship program in dentistry. You cannot legally treat a patient and experience in any meaningful way what it's like to be a dentist without, well, being in dental school or being a dentist. Now, that's for good reason. The public needs to be protected from someone performing irreversible procedures with no experience and limited or inconsistent supervision.

However, the doing of dentistry, the day-to-day actual dental procedures, makes up the vast majority of what a dentist does. Therein lies a conundrum! How do you ever really know you will enjoy doing clinical dentistry, being a dentist, if you can't ever do it prior to committing significant time and money to becoming a dentist? Maybe work in a practice as an assistant or become a dental hygienist and work in a dental office. You might at least get more up-close experience with dental procedures. You may, through your assisting or hygiene schooling, even get some hands-on experience doing limited procedures on plastic teeth.

My point is, be very, very careful jumping to conclusions that clinical dentistry is going to be easy or even fun. It may be all that for you, but it quite possibly will not. Really give it some thought. How much do you enjoy repetitive tasks? How do you handle anxious, rude, or even combative people? How do you deal with rejection? What will you do if a procedure doesn't turn out well, in spite of your best efforts?

When I went to dental school, I really had given no serious thought to how I would deal with the stresses of clinical practice. I had no experience in providing clinical care. I had not been honest with myself about how I would handle rejection, stress, and the like, and I really had no way to gauge my like or dislike of performing day-to-day dental procedures.

Take some time to be honest with yourself about your talents and skills. Be really honest with yourself about how you handle stress. Be honest with yourself about how you handle difficult situations and how well you are able to accept less than perfection. You may not be able to know for sure if clinical dentistry is for you. But by taking the time to honestly evaluate yourself and by accepting the fact that performing clinical dentistry is challenging under the best of circumstances, you can at least come to a more informed decision.

Being more honest with myself before I even applied to dental school, I may have chosen another career path. Do yourself a favor and think long and hard before you commit substantial time and serious money into a dental career. Then with your eyes wide open, give it your all and enjoy your career!

We, the authors, have wished that we would have had access to such a book prior to endeavoring on our quest to become dentists. Before dental school, many hours were spent observing and talking with dentists; however, nothing could have prepared us for what we actually experienced in dental school and what we now experience as actual dentists and business owners. We would like nothing more than to provide this as a resource for predental students or anyone considering dentistry as a profession in order to get a better idea of the career that they are about to sacrifice many years and hundreds of thousands of dollars pursuing.

Made in the USA
Columbia, SC
02 October 2020